Bookland

A TREASURY OF WINDMILL BOOKS

Edited by Pamela Kraus

Windmill Books/Simon & Schuster
New York

CONTENTS

D N S 5 X.

Boris Bad Enough

Boris was bad.

**"I hope he doesn't get any worse," said his mother.
"He'd better not," said his father. "Boris is bad enough!"**

But Boris did get worse!
And worse. And worse. And worse!

Boris's father yelled at him. Boris's mother yelled at him.
"Me bad, bad, bad? Of course!" said Boris.

"I **have** been bad," said Boris.

And he tried to be good.

"Boris could be a little better," said Boris's father.

"Nonsense," said Boris's mother. "Boris is good enough!"

There was an old woman tossed in a basket,
    Seventeen times as high as the moon;
But where she was going no mortal could tell,
    For under her arm she carried a broom.
"Old woman, old woman, old woman," said I,
    "Whither, oh whither, oh whither so high?"
"To sweep the cobwebs from the sky,
    And I'll be with you by-and-by."

There was an old woman
Lived under a hill,
And if she isn't gone,
She lives there still.

Little King Pippin he built a fine hall,
  Pie-crust and pastry-crust that was the wall.
The windows were made of black pudding and white
  And slated with pancakes, you ne'er saw the like.

14

M ommy says that she
wants a horse that is
sixteen hands.

Leo the Late Bloomer

Leo couldn't do anything right.

He couldn't read.

He couldn't write.

He couldn't draw.

He was a sloppy eater.

And, he never said a word.

"What's the matter with Leo?"
asked Leo's father.
"Nothing," said Leo's mother.
"Leo is just a late bloomer."
"Better late than never," thought Leo's father.

Every day Leo's father watched him
for signs of blooming.

And every night Leo's father watched him
for signs of blooming.

"Are you sure Leo's a bloomer?"
asked Leo's father.
"Patience," said Leo's mother,
"A watched bloomer doesn't bloom."

So Leo's father watched television
instead of Leo.

The snows came.
Leo's father wasn't watching.
But Leo still wasn't blooming.

The trees budded.
Leo's father wasn't watching.
But Leo still wasn't blooming.

Then one day,
in his own good time,
Leo bloomed!

He could read!

He could write!

He could draw!

He ate neatly!

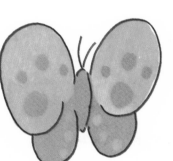

He also spoke.
And it wasn't just a word.
It was a whole sentence.
And that sentence was...

"I made it!"

I, MOUSE

I am a mouse.

Cats chase me.

People set traps to catch me.

Is it my fault I'm a mouse?

I like to munch on cheese. Is that a crime?

I like to dance with other mice.

 I like people.

But people don't like me.

I live in a big house

with a mother, father, and a little boy.

But they're always setting traps to catch me.

I'm clever, though, and they haven't caught me yet.

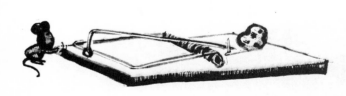
Knock on wood.

If I could only think of a way to make them like me.

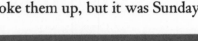

One morning, their alarm clock didn't go off, so I woke them up, but it was Sunday.

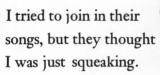

I tried to join in their songs, but they thought I was just squeaking.

I even tried to help wash the dishes, but the mother shooed me away.

I can dance just as well as that mouse on television, but the little boy prefers to watch him.

Good grief! A burglar just came through the window.

Immediately I throw myself at him and bite his ankle as hard as I can.

We fall to the floor and I stand on his chin and punch his nose.

The coward pulls a knife,

so I bite his wrist as hard as I can.

The noise of the scuffle awakens the family and they hurry downstairs.

"Please call off your mouse," cries the burglar. "He's a terror."

"Good mouse," says the mother. "Brave mouse," says the father. "Watchmouse!" says the little boy.

The little boy shakes my hand. I am so proud and happy.

The burglar says he is sorry. But as he broke the law he is sent to jail anyway.

They put my picture in all the papers and I receive thirty-eight fan letters, all of which I answer personally.

WATCHMOUSE THUMPS THUG

MOUSE HERO

BEWARE OF MOUSE

WATCHMOUSE

Best of all the family loves me now, and that, with a little cheese, is really all a mouse's heart desires.

SONGS
from
ROLAND
the Minstrel Pig

Hail the heavens up above,
Hail to honor, courage, love.
Let these halls with music ring!
All hail His Majesty the King!

Hey, nonny, nonny, this weather is bonny,
The flowers are all in bloom.
Let us be gay as we go on our way
Singing ta-ra, ta-ra-ra, ta-boom.

Farewell, dear world, dear hill, dear shore,
Dear butterflies, dear birds, dear bees,
Dear night, dear day, dear seasons four,
Dear flowering fields, dear fruited trees,
Dear warming sun, dear gentle breeze.
My heart's so sore
I'll be no more.
I feel an aching in my knees.

Lonely am I as yonder moon
That roams the empty sky.
No one's here to hear my tune.
I'm sad enough to cry.

I loved her tail, her ears, her snout,
I loved her form so fair.
She looked at me with glowing eyes;
I walked as if on air.

23

Dance to your partner—one, two, three—
With merry leaps and gracious bends.

Life is a lark. With spirits free
We'll frisk about till the evening ends.

MILTON THE EARLY RISER

Milton woke up early...
and went out to play.
But there was nobody to play with.

The Creeps next door were still asleep.
So were the Whippersnappers across the way.
and the Nincompoops in the back.
The whole world seemed to be asleep.

So Milton watched television.
And the sleepers slept on.

Milton jumped up and down.
But the sleepers slept on.

Milton danced and did tricks.
Yet the sleepers slept on.

Then Milton sang up a storm.
The mountains shook.
The trees trembled.
And a whirlwind blew the sleepers out of bed!
Still the sleepers slept on.

"Oh dear," said Milton. "What a mess."

And he worked and worked
and he put things right.

Just as everybody woke up.

Everybody but Milton.

"Rise and shine," said Milton's father.
"Wake up sleepy-head," said Milton's mother.

Milton the early riser didn't hear a word.
He was fast asleep.

29

I'm a monkey.

My mother is a monkey.

My father is a monkey.

My brothers are monkeys.

My sisters are monkeys.

My grandmother is a monkey.

My grandfather is a monkey.

My aunt is a monkey.

My uncle is a monkey.

My great-aunt is a monkey.

My great-uncle is a monkey.

My cousins are monkeys.

I have a best friend...

...who lives next door.

He's a monkey too.

Daddy says there was a king
who rained for forty years.

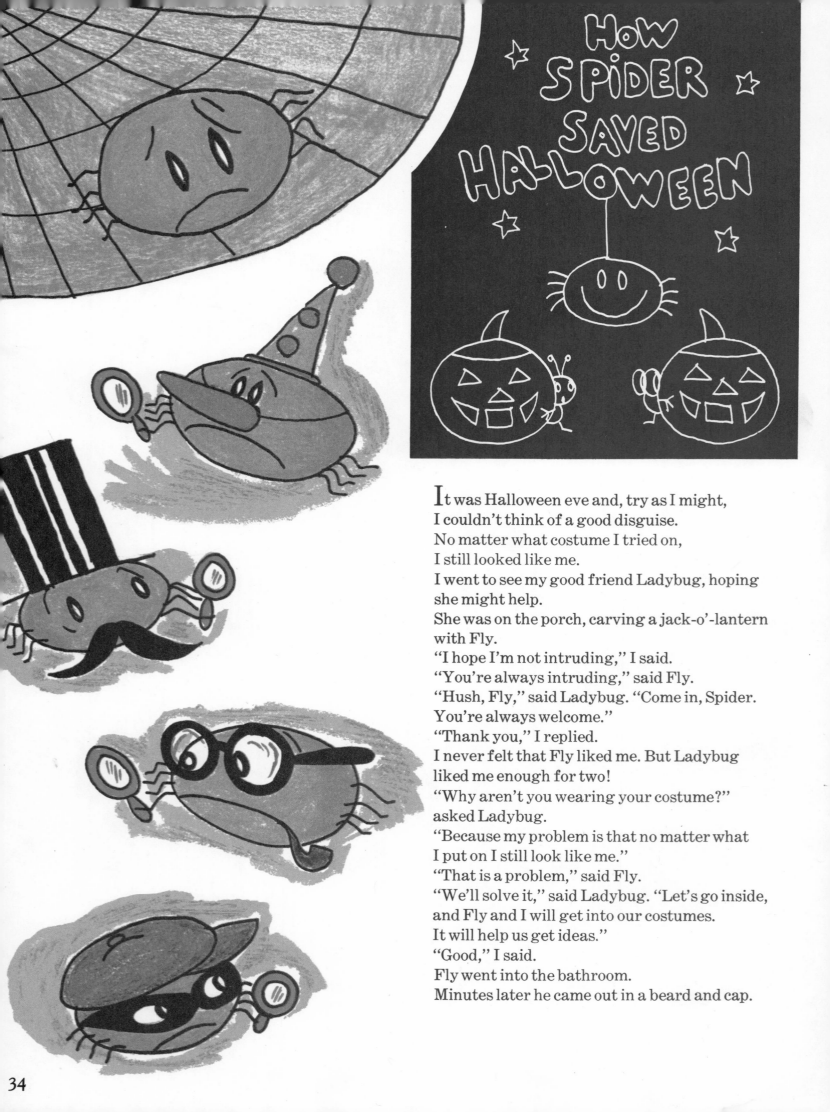

How SPIDER SAVED HALLOWEEN

It was Halloween eve and, try as I might,
I couldn't think of a good disguise.
No matter what costume I tried on,
I still looked like me.
I went to see my good friend Ladybug, hoping
she might help.
She was on the porch, carving a jack-o'-lantern
with Fly.
"I hope I'm not intruding," I said.
"You're always intruding," said Fly.
"Hush, Fly," said Ladybug. "Come in, Spider.
You're always welcome."
"Thank you," I replied.
I never felt that Fly liked me. But Ladybug
liked me enough for two!
"Why aren't you wearing your costume?"
asked Ladybug.
"Because my problem is that no matter what
I put on I still look like me."
"That is a problem," said Fly.
"We'll solve it," said Ladybug. "Let's go inside,
and Fly and I will get into our costumes.
It will help us get ideas."
"Good," I said.
Fly went into the bathroom.
Minutes later he came out in a beard and cap.

"Who am I?" asked Fly.

"Santa Claus," I replied.

"Wrong," said Fly. "I'm one of the seven dwarfs."

Then Ladybug excused herself and went into
the bathroom to change into her costume.
Minutes later she emerged.

"What am I?" asked Ladybug.

"A witch," I said.

"Correct," said Ladybug.

"Now for you," said Ladybug.

She dressed me up in ladies' clothes.

"You still look like you," said Fly.

"Still looks like himself," said Ladybug.

"Looks even more like himself," said Fly.

A cowboy suit wasn't much better.

We all put our heads together to think of a
costume that would *really* disguise me. Suddenly
there was a sickening *SQUASH* from the porch.
We all ran outside to see what it was.
Ladybug's jack-o'-lantern was wrecked!
Two bullies with baseball bats were hooting
and running away.

Ladybug burst into tears. "Our beautiful pumpkin!"
she cried.

"And it's too late to get another," moaned Fly.
"Halloween is ruined!"

"It's true," I said. "Halloween without a pumpkin
just isn't Halloween."

I stopped feeling sorry for myself and started
feeling sorry for my friends.

Then I got an idea.

"Color me orange!" I said.

They colored me orange.

While they were coloring me orange,
I cut up a piece of green construction paper.

"Paste this on my head," I said.

Fly got the paste jar.

Ladybug did the pasting.

Then I took a black marker and drew lines on
myself and blacked out my front teeth.

"You're a pumpkin!" they both exclaimed.
"Halloween isn't spoiled after all!"

"You're darn tootin'" I said. "On to trick or treat!"

We tricked-or-treated all over the neighborhood,
and everyone gave us lots of treats and
admired our costumes. Especially the walking,
talking pumpkin.

At last our bags were full
and we headed home.
When who should we see coming
down the street but the two bullies
who had smashed the jack-o'-lantern. They were heading
our way with cans of shaving cream.
"They'll spray us with shaving cream!" cried Ladybug.
"They'll steal our trick or treats," moaned Fly.
"No they won't," I said. "They're bullies, and bullies
are cowards. Quick. Hide behind this bush."
Just as the bullies passed by,
I jumped out
and screamed, "Boo!"
The bullies dropped their shaving cream
and ran away screaming, "A pumpkin ghost!
A pumpkin ghost! Save us! Save us!"
We went back to Ladybug's house and counted
our loot. Ladybug planted a kiss on my cheek and
Fly shook my hand.
"Because you saved Halloween," said Ladybug.
"I must admit you did it," said Fly. "You saved
Halloween."
I guess I had. And I was very happy to have saved
Halloween for my two dear friends, Fly and Ladybug.

The End

C D B !

D B S A B-Z B.
O, S N-D !

Sylvester
and the
Magic Pebble

Sylvester Duncan lived with his mother and father at Acorn Road in Oatsdale. One of his hobbies was collecting pebbles of unusual shape and color.

On a rainy Saturday during the holidays he found a quite extraordinary one. It was flaming red, shiny, and perfectly round, like a marble. As he was studying this remarkable pebble, he began to shiver, probably from excitement, and the rain felt cold on his back. "I wish it would stop raining," he said.

To his great surprise the rain stopped. It didn't stop gradually as rains usually do. It CEASED. The drops vanished on the way down, the clouds disappeared, everything was dry, and the sun was shining as if rain had never existed.

In all his young life Sylvester had never had a wish gratified so quickly. It

struck him that magic must be at work, and he guessed that the magic must be in the remarkable-looking red pebble. (Where indeed it was.) To make a test, he put the pebble on the ground and said, "I wish it would rain again." Nothing happened. But when he said the same thing holding the pebble in his hoof, the sky turned black, there was lightning and a clap of thunder, and the rain came shooting down.

He wished the sunshine back in the sky, and he wished a wart on his left hind fetlock would disappear, and it did, and he started home, eager to amaze his father and mother with his magic pebble. He could hardly wait to see their faces. Maybe they wouldn't even believe him at first.

"What a lucky day this is!" thought Sylvester. "From now on I can have anything I want. My father and mother can have anything they want. My relatives, my friends, and anybody at all can have everything anybody wants!"

As he was crossing Strawberry Hill, thinking of some of the many, many things he could wish for, he was startled to see a mean, hungry lion looking right at him from behind some tall grass. He was frightened. If he hadn't been so frightened, he could have made the lion disappear, or he could have wished himself safe at home with his father and mother.

He could have wished the lion would turn into a butterfly or a daisy or a gnat. He could have wished many things, but he panicked and couldn't think carefully.

"I wish I were a rock," he said, and he became a rock.

The lion came bounding over, sniffed the rock a hundred times, walked around and around it, and went away confused, perplexed, puzzled, and bewildered. "I saw that little donkey as clear as day. Maybe I'm going crazy," he muttered.

And there was Sylvester, a rock on Strawberry Hill, with the magic pebble lying right beside him on the ground, and he was unable to pick it up. "Oh, how I wish I were myself again," he thought, but nothing happened. He had to be touching the pebble to make the magic work, but there was nothing he could do about it.

His thoughts began to race like mad. He was scared and worried. Being helpless, he felt hopeless. He imagined all the possibilities, and eventually he realized that his only chance of becoming himself again was for someone to

find the red pebble and to wish that the rock next to it would be a donkey. Someone would surely find the red pebble—it was so bright and shiny—but what on earth would make them wish that a rock were a donkey? The chance was one in a billion at best.

Sylvester fell asleep. What else could he do? Night came with many stars.

Meanwhile, back at home, Mr. and Mrs. Duncan paced the floor, frantic with worry. Sylvester had never come home later than dinner time. Where could he be? They stayed up all night wondering what had happened, expecting that Sylvester would surely turn up by morning. But he didn't, of course. Mrs. Duncan cried a lot and Mr. Duncan did his best to soothe her. Both longed to have their dear son with them.

"I will never scold Sylvester again as long as I live," said Mrs. Duncan, "no matter what he does."

At dawn, they went about inquiring of all the neighbors.

They talked to all the children—the puppies, the kittens, the colts, the piglets. No one had seen Sylvester since the day before yesterday.

They went to the police. The police could not find their child.

All the dogs in Oatsdale went searching for him. They sniffed behind every rock and tree and blade of grass, into every nook and gully of the neighborhood and beyond, but found not a scent of him. They sniffed the rock on Strawberry Hill, but it smelled like a rock. It didn't smell like Sylvester.

After a month of searching the same places over and over again, Mr. and Mrs. Duncan no longer knew what to do. They concluded that something dreadful must have happened and that they would probably never see their son again. (Though all the time he was less than a mile away.)

They tried their best to be happy, to go about their usual ways. But their usual ways included Sylvester and they were always reminded of him. They were miserable. Life had no meaning for them any more.

Night followed day and day followed night over and over again. Sylvester on

the hill woke up less and less often. When he was awake, he was only hopeless and unhappy. He felt he would be a rock forever and he tried to get used to it. He went into an endless sleep. The days grew colder. Fall came with the leaves changing color. Then the leaves fell and the grass bent to the ground.

Then it was winter. The winds blew, this way and that. It snowed. Mostly, the animals stayed indoors, living on the food they had stored up.

One day a wolf sat on the rock that was Sylvester and howled and howled because he was hungry.

Then the snows melted. The earth

warmed up in the spring sun and things budded.

Leaves were on the trees again. Flowers showed their young faces.

One day in May, Mr. Duncan insisted that his wife go with him on a picnic "Let's cheer up," he said. "Let us try to live again and be happy even though Sylvester, our angel, is no longer with us." They went to Strawberry Hill.

Mrs. Duncan sat down on the rock.

The warmth of his own mother sitting on him woke Sylvester up from his deep winter sleep. How he wanted to shout, "Mother! Father! It's me, Sylvester, I'm right here!" But he couldn't talk. He had no voice. He was stone-dumb.

Mr. Duncan walked aimlessly about while Mrs. Duncan set out the picnic food on the rock—grass sandwiches, pickled oats, spinach salad, thistle tart. Suddenly Mr. Duncan saw the red pebble. "What a fantastic pebble!" he exclaimed. "Sylvester would have loved it for his collection." He put the pebble on the rock.

They sat down to eat. Sylvester was now as wide awake as a donkey that was a rock could possibly be. Mrs. Duncan felt some mysterious excitement. "You know, Father," she said suddenly. "I have the strangest feeling that our dear Sylvester is still alive and not far away."

"I am, I am!" Sylvester wanted to shout, but he couldn't. If only he had realized that the pebble resting on his back was the magic pebble!

"Oh, how I wish he were here with us on his lovely May day," said Mrs. Duncan. Mr. Duncan looked sadly at the ground. "Don't you wish it too, Father?" she said. He looked at her as if to say, "How can you ask such a question?"

Mr. and Mrs. Duncan looked at each other with great sorrow.

"I wish I were myself again, I wish I were my real self again!" thought Sylvester.

And in less than an instant, he was!

You can imagine the scene that followed—the embraces, the kisses, the questions, the answers, the loving looks, and the fond exclamations!

When they had eventually calmed down a bit, and had gotten home, Mr. Duncan put the magic pebble in an iron safe. Some day they might want to use it, but really, for now, what more could they wish for? They all had all that they wanted.

JUNIOR THE SPOILED CAT

Junior was a spoiled cat.
His Mother didn't spoil him.
His Father didn't spoil him.
Billy spoiled him.

Billy dangled a string
in front of him for hours.

Billy fed him
at the table.

Billy played Peek-A-Boo with him.
Billy even stood on his head
to amuse Junior.
And at night, Junior slept on
Billy's head.

"The things that boy does for
that cat!" Billy's Mother said.
"And that cat doesn't even
appreciate it!" Billy's Father said.
"That Junior is a spoiled cat!"
they both said.

Then one day Billy felt sick.
His parents put him to bed
and called the doctor.

Dr. Green arrived
and leaped up the stairs
to Billy's room.

He gave Billy a thorough
examination and declared,
"You are suffering from the
common tummy ache!"
"Keep him quiet—and keep
this cat out of his room,"
said the doctor, who didn't
like cats in rooms even
when people were well.

Billy's parents kept his door shut and Junior sat outside the door.

And Billy kept getting sicker

and sicker

and sicker.

Dr. Green got excited and gave Billy all sorts of medicines. He called in all sorts of special doctors.

But the best thing he did was to leave the door open.
And in walked Junior.
Junior jumped up on Billy's bed to the horror of the doctors.

He licked Billy's pale little face. Billy opened his eyes. "Junior," he said, "I sure missed you." He gave Junior a hug and Junior began to purr. The more Billy hugged the cat the less sick he looked.

Until soon

he looked like himself again. The doctors were amazed!

"That spoiled cat saved our son's life!" said Billy's parents. "From now on we'll spoil him, too."

But Junior ignored them.

He only lets Billy spoil him.

Because it's not that he likes being spoiled—

He just likes Billy!

"I can't sleep," said Richard Rabbit.
"You must try," said his mother.

"But there is a giant walking around downstairs," said Richard Rabbit.

"Silly," said his mother.
"That is only your father."

"There are eight little pigs in my bed,"
said Richard.

"Those are your toes," said Mother.

"There is a man looking in the window," said Richard.

"That is only the Man in the Moon,"
said Mother. "Now, go to sleep."

"There are rocks in my bed,"
said Richard.

"Nonsense," said Mother.
"Your sheets need smoothing."

Then the clock struck six and Herman
  hurried home.
He washed his hands and face.
And sat down to supper.

"May I help you to some mashed potatoes?"
  asked Herman's father.
"No thanks," said Herman, "I'll help
  myself."

The Brownies' Joke Book

Palmer Cox

Text by Robert and Pam Kraus
Color collages by Pam Kraus

A laugh a day keeps the doctor away.

"Knock, knock. Are you out?"
"Gnome home."

Q. Why did the fairy thumb his nose at the Brownie?
A. He had no Elf-respect.

"You ought to eat more. You're just bones and bones."

Q. What do you call a Brownie accountant?
A. A pocket calculator.

Q. What is the Brownies' favorite movie?
A. Snow White and the Seven Brownies.

Q. What did the mean Brownie king yell?
A. "Dwarf with his head!"

Q. Why does a Brownie wear pink suspenders?
A. To keep his pants up.

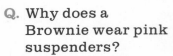

Q. What do you call a playboy Brownie?
A. A Brunnie.

"There's a fly in my soup."
"No extra charge for the fly."

Q. What did the Brownie do on New Year's Eve?
A. Got pixilated.

Q. What is the Brownies' favorite song?
A. "Gnome, Sweet Gnome."

Now you're really getting my goat.

Q. Why did the two Brownies kiss under the toadstool?

A. They thought it was a mushroom.

I wear top hats
I wear tails,
You won't find
My kind in jails.

"Knock, knock."
"Who's there?"
"Goblin."
"Goblin who?"
"Goblin my dinner,
Go away!"

Q. Why did the Brownie cross the road?

A. To catch the chicken.

Q. What do you call a day-old Brownie?

A. Stale.

Q. Why did the Elf wear chocolate trousers?

A. He wanted a brown knee.

Q. Where does Betty Brownie live?

A. In Gnome, I think, but Alaska.

Q. What did the Brownie say to the Leprechaun?

A. That's enough of your lep.

"Brownie can you spare a dime?"

I'm a Brownie,
Plump and furry.
My friends call me
Furry Murray.

"Are you sure that's the tooth, the whole tooth, and nothing but the tooth?"

Th-Th-That's all the jokes, folks.

51

DOG FOR SALE

Jared bought a dog.

He called the dog Charlie.

Charlie fell sick and Jared looked after him
well. The next day Jared got a big surprise.

Charlie should have been named Arlene.

Did you ever hear such a bunch of fairy tails?